Contents

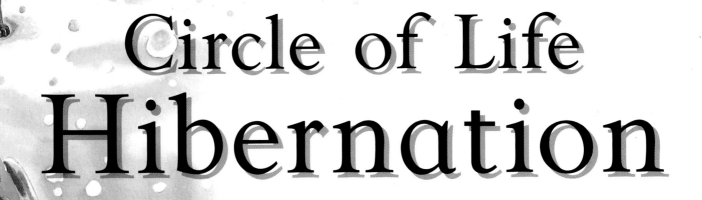

Circle of Life
Hibernation

Written and illustrated by Carolyn Scrace
Created and designed by David Salariya

W
FRANKLIN WATTS
LONDON•SYDNEY

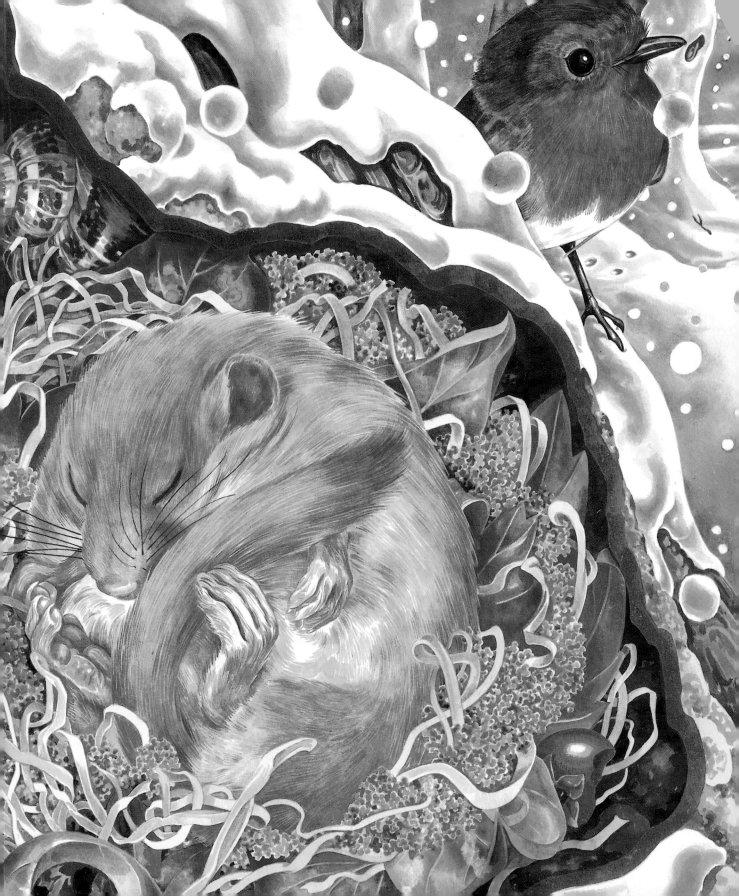

Introduction

When it is cold animals cannot find much food. Some go to sleep for the whole of winter.

This is called hibernation.

In the months before they hibernate, the animals eat as much as possible and get fat. During hibernation, their bodies use up this fat. They get very cold and thin. Their breathing and heartbeat also slow down.

In the warm spring weather the hibernating animals wake up again.

Bears, bats, squirrels and dormice are all animals that hibernate.

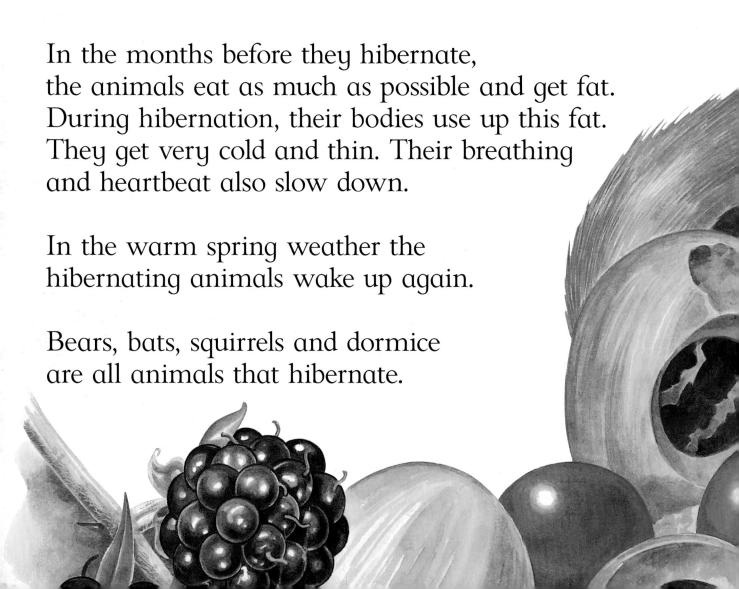

Fox

Squirrel

Rabbits

Finding food

In early autumn, the dormouse can find plenty of food.

It eats nuts, seeds, fruit, bark and shoots as well as snails and insects.

Hawthorn berries

Snail

Hazelnut

Eating hazelnuts

The dormouse likes the taste of hazelnuts.

First it gnaws a hole in the shell with its strong lower teeth.

Then it eats the kernel inside.

The dormouse eats so many hazelnuts that it becomes very fat.

Blackberry

9

Leaves

Digging a hole

Later in the autumn
there is less food for the
dormouse to eat.

It then starts digging
a hole under leaves
or the roots of a tree.

In the autumn, the days
get shorter and the
weather gets colder.
Leaves start to fall
off the trees and
many plants die.

Earthworm

nail

Ready to build a nest

The dormouse has finished digging its hole.

The hole is just the right size for the dormouse to make a nest in.

The snail is also looking for shelter from the cold.

13

15

Making the nest comfortable

The dormouse uses moss, leaves and grass to make its nest warm and comfortable.

It will spend 5 months hibernating inside the nest.

The magpie has found a juicy worm to eat.

Magpie

...use
...ntrance
... with
...eaves.

...nimals such as
...and badgers
will eat the
dormouse
if they
find it.

Moss

Getting ready for hibernation

In its nest the dormouse settles down for the winter.

In case it wakes up for a short time, it has put a small store of nuts and berries in the nest.

During hibernation the dormouse gets much thinner. Its heartbeat and breathing slow down and its body gets colder.

Berries

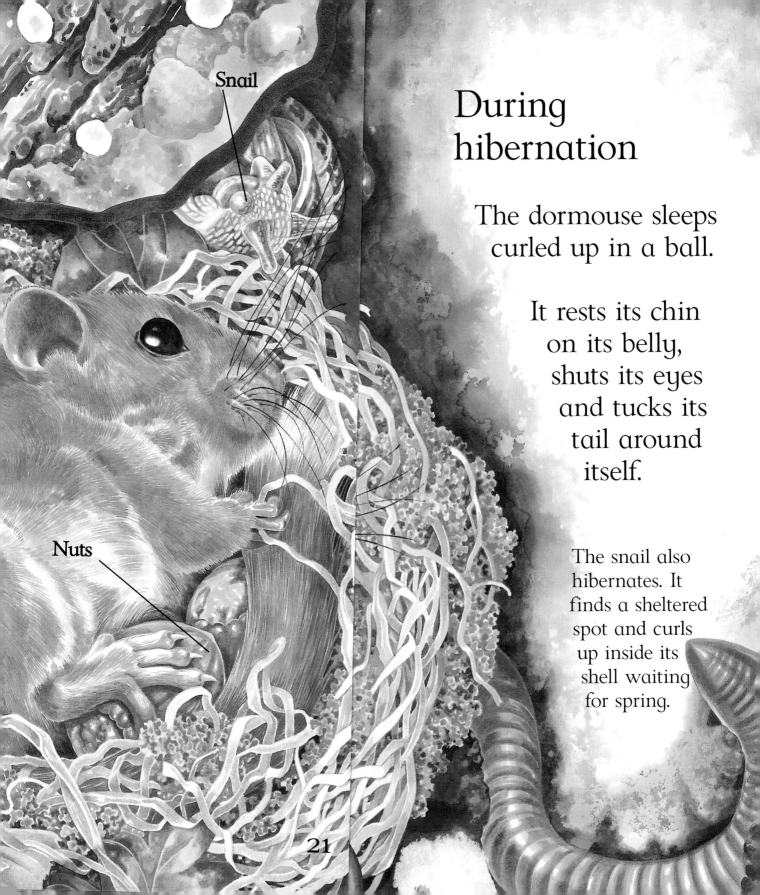

Snail

Nuts

During hibernation

The dormouse sleeps curled up in a ball.

It rests its chin on its belly, shuts its eyes and tucks its tail around itself.

The snail also hibernates. It finds a sheltered spot and curls up inside its shell waiting for spring.

Waking up

In the spring it gets warmer. The dormouse wakes up.

It takes the little animal twelve hours to wake from its hibernation.

In spring, the trees grow new leaves and shoots. Spring flowers begin to grow.

Magpies

Rabbits

Bluebells

Snail

...ving
...nest

...rmouse has
...d the long, cold
...months.

...now very
...gry. It climbs
...of its nest and
...es in search of
food again.

As the earth
warms up, the
snail also
wakes up
and comes
out of its shell.

27

The dormouse's year

During summer and early autumn, the dormouse finds lots of food to eat.

In late autumn the dormouse picks a safe place to dig a hole for its nest.

The dormouse uses bits of grass, leaves and moss to build its nest.

28

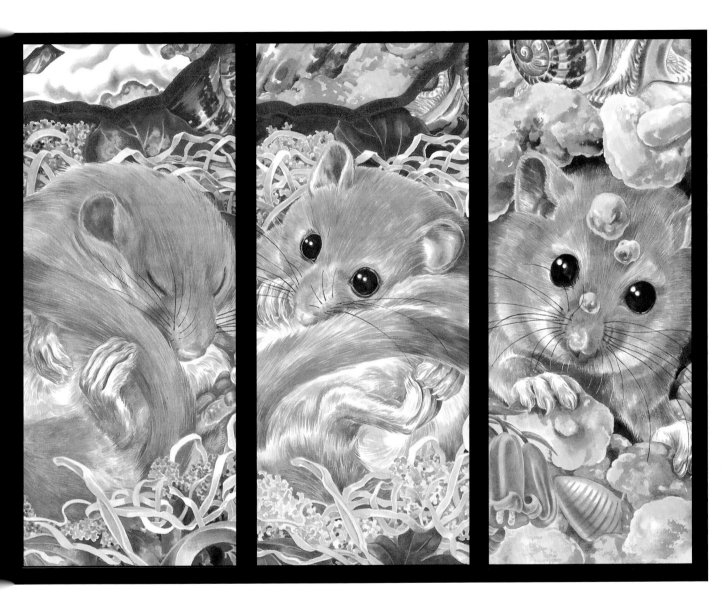

From late autumn until the middle of spring, the dormouse stays in hibernation.

5 to 6 months later, in the middle of spring, the dormouse wakes up.

The dormouse leaves its nest and goes in search of food. It starts to eat again.

Hibernation words

Autumn
The months of September, October and November, when the weather gets cold.

Bark
The tough outer covering of a tree.

Earthworm
An animal with a long thin body divided into small parts called segments, and no bones. It lives in the soil.

Hazelnut
The small fruit or nut of the hazel tree.

Insect
An animal that has a hard outer covering and a body divided into 3 parts: the head, the thorax and the abdomen. Insects have 6 legs.

Kernel
The part of a nut which can be eaten. It is found inside the hard shell.

Moss
A small green plant that grows on damp soil and on some trees and stones.

Nest
A bed or shelter made by an animal.

Shoots
The new, young growth on plants and trees.

Snail
An animal that does not have any bones. It has a hard spiralled shell and a soft body.

Spring
The months of March, April and May, when the weather begins to get warmer.

Winter
The months of December, January and February, when the weather is very cold.

Index

31

Created, designed and produced by
The Salariya Book Company Ltd
Book House,
25 Marlborough Place,
Brighton BN1 1UB

Visit the Salariya Book Company at www.salariya.com

A CIP catalogue record for this book is available from
the British Library.

ISBN 0 7496 4425 7

This edition first published in 2005 by
Franklin Watts
96 Leonard Street
London
EC2A 4XD

Franklin Watts Australia
45-51 Huntley Street
Alexandria
NSW 2015

Printed in Hong Kong

Natural History Consultant   Dr Gerald Legg
Language Consultant   Betty Root

Editor   Karen Barker Smith
Assistant Editor   Stephanie Cole